Five Fatality Residential Motel Fire
Thornton, Colorado

Investigated by: Thomas H. Miller, P.E.

This is Report 104 of the Major Fires Investigation Project conducted by Varley-Campbell and Associates, Inc./TriData Corporation under contract EMW-94-4423 to the United States Fire Administration, Federal Emergency Management Agency.

Department of Homeland Security
United States Fire Administration
National Fire Data Center

U.S. Fire Administration Fire Investigations Program

The U.S. Fire Administration develops reports on selected major fires throughout the country. The fires usually involve multiple deaths or a large loss of property. But the primary criterion for deciding to do a report is whether it will result in significant "lessons learned." In some cases these lessons bring to light new knowledge about fire--the effect of building construction or contents, human behavior in fire, etc. In other cases, the lessons are not new but are serious enough to highlight once again, with yet another fire tragedy report. In some cases, special reports are developed to discuss events, drills, or new technologies which are of interest to the fire service.

The reports are sent to fire magazines and are distributed at National and Regional fire meetings. The International Association of Fire Chiefs assists the USFA in disseminating the findings throughout the fire service. On a continuing basis the reports are available on request from the USFA; announcements of their availability are published widely in fire journals and newsletters.

This body of work provides detailed information on the nature of the fire problem for policymakers who must decide on allocations of resources between fire and other pressing problems, and within the fire service to improve codes and code enforcement, training, public fire education, building technology, and other related areas.

The Fire Administration, which has no regulatory authority, sends an experienced fire investigator into a community after a major incident only after having conferred with the local fire authorities to insure that the assistance and presence of the USFA would be supportive and would in no way interfere with any review of the incident they are themselves conducting. The intent is not to arrive during the event or even immediately after, but rather after the dust settles, so that a complete and objective review of all the important aspects of the incident can be made. Local authorities review the USFA's report while it is in draft. The USFA investigator or team is available to local authorities should they wish to request technical assistance for their own investigation.

This report and its recommendations were developed by USFA staff and by Varley-Campbell & Associates, Inc. Miami and Chicago, its staff and consultants, who are under contract to assist the Fire Administration in carrying out the Fire Reports Program.

The U.S. Fire Administration greatly appreciates the cooperation received from Fire Marshall Steven Pischke of the North Metro Fire Rescue Authority, Detective Gerald Staubinger of the Thornton Police, and Special Agent Deborah Dassler of the Bureau of Alcohol, Tobacco & Firearms in preparing this report.

For additional copies of this report write to the U.S. Fire Administration, 16825 South Seton Avenue, Emmitsburg, Maryland 21727. The report is available on the Administration's Web site at http://www.usfa.dhs.gov/

U.S. Fire Administration
Mission Statement

As an entity of the Department of Homeland Security, the mission of the USFA is to reduce life and economic losses due to fire and related emergencies, through leadership, advocacy, coordination, and support. We serve the Nation independently, in coordination with other Federal agencies, and in partnership with fire protection and emergency service communities. With a commitment to excellence, we provide public education, training, technology, and data initiatives.

TABLE OF CONTENTS

Five Fatality Residential Motel Fire
11th East 84th Avenue
Thornton, Colorado
January 27, 1997

Investigated by: Thomas H. Miller, P.E.

Local Contact: Chief John J. O'Hayre
                    Fire Marshal Steven J. Pischke
                    Fire Safety Engineer A. Keith Brown
                    North Metro Fire Rescue Authority
                    10550 Huron St.
                    Northglenn, Colorado 80234

                    Detective Gerald Straubinger
                    Thornton Police
                    9500 Civic Center Drive
                    Thornton, Colorado 80229

                    Special Agent Deborah J. Dassler
                    Bureau of Alcohol, Tobacco & Firearms
                    1700 Broadway, Suite 1010
                    Denver, Colorado 80290

OVERVIEW

At 2:33 a.m., The Adams County Communications Center (ADCOM) received a telephone report of smoke and fire at the Hacienda Plaza Inn. Police officers who reached the scene within two to three minutes of the alarm reported flames visible at the exterior and occupants evacuating into the 17 degree night air.

Four occupants were trapped in two separate rooms and perished. A fifth fatality, who had been staying in one of these two rooms, was later discovered at the west exit from the second floor breezeway. Two firefighters sustained minor injuries while fighting the fire.

The cause of the fire has been determined to be arson by persons unknown. A $15,000.00 reward for information leading to the arrest and conviction of the responsible person or persons has been advertised by local television, radio, and print media in the Denver area. The Bureau of Alcohol,

1

Tobacco & Firearms activated a National Response Team at the request of local officials to assist with the scene investigation and interviewing of witnesses.

The fatalities were staying in rooms that did not exit directly to the outside of the building. Rather, these rooms discharged through an enclosed breezeway area which contained two open stairs that connected three levels of the motel. Investigators believe that the point or origin was near the bottom of the stairway on the lowest level. The smoke, heat, and fire spread up this stairway, rapidly engulfing the breezeway interior.

The section of the motel where the fire originated was not provided with automatic sprinklers. The first notification of the fire was from Room 222 when one of the occupants telephoned the motel's front desk to report smoke entering the room from the bathroom exhaust. Both occupants of this room perished in the fire.

The fire department estimated the fire loss at $800,000 to the property and contents. The motel is currently being prepared for purchase and demolition by the city as part of a redevelopment project in the area. The city had been interested in the property for a number of years before the fire and had been discussing its purchase with the owner. Overall, the property was in good condition for its age and repairs had not been neglected.

KEY ISSUES

Issues	Comments
"Center Loaded" Corridors	All of the fatalities were either found in a room or occupied a room which did not open directly to the outside. All occupants whose rooms exited to the outside covered walkways escaped.
Lack of Automatic Sprinklers	The fire's point of origin was in an area that was not provided with automatic sprinklers. The fire developed in the closet releasing heat and partially burned gases into the stairway where they ignited on the lower and upper levels.
Lack of a Building-Wide Fire Detection and Alarm System	Most motel occupants were asleep at the time of the fire. Although each motel room had a battery operated single station smoke detector, there were no detectors in common areas, storage rooms and utility spaces located outside the main lobby section. There was no alarm system to notify occupants to evacuate in the section of the building where the fire started.
Combustible Concealed Space	A two story high combustible utility space connected the rooms in each wing together, both horizontally and vertically. This space contained plumbing, electrical, cable TV, and toilet exhaust ducts. It provided a ready avenue for smoke and fire to spread.
Pre-incident Planning	The fire department prepared a pre-incident plan several years before the fire showing room numbers, stairways, hydrants and major building features. Unfortunately, the motel owners had reorganized the room numbers shortly before the fire. This resulted in some confusion early in the incident regarding the occupied versus unoccupied rooms.
Unprotected Vertical Opening	The fire is believed to have started on the lowest level with smoke, heat, and fire spreading up the open stairway into the enclosed corridor, trapping occupants in their rooms.

FIRE DEPARTMENT

North Metro Fire Rescue Authority was created on January 1, 1994 by the merger of the Thornton Fire Department and West Adams County Fire Protection District. The Authority serves approximately 126 square miles with nearly 136,000 people in a growth area north of Denver. It is staffed by a combination of career and volunteer firefighters operating out of eight fire stations on a three-shift schedule. A separate headquarters facility provides offices, training room, and repair/maintenance garage.

The Authority operates eight engine companies, a brush truck, a special rescue (for low angle and trench), and two reserve engines. In addition, three private ambulance companies provide ambulance transport units for the Authority. Career staffing on each shift consists of 29 total personnel led by two battalion chiefs, seven or eight station officers, and nineteen or twenty paramedics, engineers, and firefighters. Approximately 40 volunteers are active with the Authority and completely staff one of the eight stations.

Headquarters includes fire chief, deputy chief, division chiefs for training and prevention, volunteer coordinator, finance manager, and administrative services. The Authority supports active public education efforts and fire prevention/plans review through headquarters staff. The fire marshal is also responsible for investigation of fires with the assistance of city and county law enforcement members.

Dispatching is provided by the Adams County Communications Center (ADCOM) which handles 9-1-1 emergency calls for police and fire. They also provide the link to the private ambulance services.

BUILDING DESCRIPTION

The Hacienda Plaza Inn was a partial "U" shaped structure consisting of three wings and was constructed in 1963. (See Figure 1 for diagram). The south or front wing and the west wing contained most of the guest rooms. The first floor of the front wing had several areas which were available for office and commercial use including one which was occupied by a quick print shop. The east wing contained the registration desk, offices, lounge, restaurant, banquet room, and on the second floor, efficiency style guest rooms with small kitchens.

The front and west wings were each two stories high and had essentially the same construction. However, due to ground elevation differences, there were actually three levels of guest rooms within these two wings. (See photographs 1 to 3.) The first floor was concrete on grade and the exterior walls were masonry (brick).

On the first floor of each wing, parallel masonry block walls formed the lower part of a two story utility chase between back to back guest rooms. These block walls supported the second floor in the middle of the building. This floor consisted of concrete on Tectum board[1] which was supported by 4 x 12-inch wood beams about 34 inches on center. (See photograph 4.) The beams rested on the chase and outside walls.

---

[1] A brand name product for structural cement fiber units which are made from proprietary mixes of processed wood fibers and cementious binder. This mixture is formed into slabs or boards of various sizes and thicknesses before hardening. Uncoated boards typically have a flame spread index of about 25.

On the second floor, the utility chase was constructed of parallel 2 x 4-inch wood stud walls covered only on the guest room side only with drywall. (See photograph 5). The chase walls formed the center support for the 4 x 12-inch wood roof beams which were spaced about 34 inches on center. The beams held up a light weight roof consisting of Tectum board[1] covered with a hot mopped asphalt built-up gravel topped roof covering. (See photograph 6.)

The east wing was two stories and basement with masonry outside walls and similar construction as the front and west wings. The basement level contained a large banquet room which was protected by automatic sprinklers. The first floor contained the lobby, registration desk, lounge and restaurant. On the second floor were seventeen efficiency-style guest rooms along a center-loaded corridor and two small meeting rooms.

A typical guest room in the front and west wings was 12 feet wide and nearly 22 feet deep. A solid core wood entrance door, window and small under window, electric air conditioning and heating unit were in the wall facing the exterior walkway. For Rooms 220, 221, and 222, this wall and door opened into the enclosed breezeway. (See Figure 2 for diagram) The separation walls between adjacent guest rooms were 2 x 4-inch wood studs on 16 inch centers covered on each side with one layer. Both 1/2-inch- or 5/8-inch-thick drywall was randomly used in different room separation walls. These walls were extended from the outside wall to the utility chase wall and from the floor to either the wood beam or Tectum board above.

The roughly 5-foot-square bathroom was located in the corner of the guest room along with the utility chase wall. It contained a commode and combination shower/tub. Each bathroom had a metal toilet exhaust duct which passed into the utility chase where it was grouped with other guest room toilet exhausts to a roof-mounted fan. On the utility chase wall adjacent to the bathroom was a counter with sink.

The utility chase in the center of the front and west wings ran the entire length of each wing. The two story high chase (ground floor to the roof) was about 3 feet wide. There was no separation between floors inside the chase. (See photograph 7.) In addition, the separation between the utility chase and the ground floor guest rooms was not complete and permitted fire travel into several rooms. The utility chase contained the water, sewer toilet exhaust ducts, cable TV, and electrical wiring that supplied the guest rooms.

Most of the electrical wiring was in conduit, although nonmetallic cable ("Romex") was observed in parts of the utility chase. Overcurrent protection was provided by circuit breakers. The electrical system is not believed to be involved with the cause or spread of this fire.

Fire Protection Systems

The basement banquet room in the east wing was protected by automatic sprinklers supplied by a 6-inch private water main. The private water main ran on the west and north sides of the property. It was connected to an 8-inch city water main on the south side of 84th Street. The private main also supplied two fire hydrants that were located on the hotel property. One was located at the southwest corner of the front wing and the second was at the northeast corner of the west wing. The property did not have a standpipe system.

Each guest room contained a battery-operated single station smoke detector. The front and west wings did not have a fire alarm system that would sound throughout these wings to alert the occupants of an emergency. There was no centralized means to report a fire other than using the room telephone to notify the front desk from these two wings.

The east wing had a local fire detection and alarm system in addition to the guest room single station smoke detectors. The local system consisted of manual pull stations at the exits and smoke detectors in the corridors. Evacuation horns were located throughout this wing.

Multipurpose dry chemical fire extinguishers were distributed in cabinets throughout the property. A kitchen extinguishing system protected the cooking equipment and exhaust system in the restaurant.

THE FIRE

On the night of the fire, the hotel had guests in 72 out of 109 rooms according to records available to fire investigators. These records also account for 162 occupants of various ages in these rooms. One of the common challenges in transient occupancies such as hotels is identifying the occupied rooms and the actual number of occupants. A complication encountered by firefighters in this incident was the hotel had recently modified its room numbering system and the numbers were different from the department's preplan materials. This made initial efforts at directing crews to occupied rooms confusing. Eventually, all rooms were searched beginning with those closest to the visible fire to assure each room was physically checked.

Around 2:30 a.m. one of the occupants of Room 222 telephoned the front desk to report smoke coming through the bathroom exhaust. It is unknown if she was awake when the smoke was discovered or if she was awaked by the room's single station smoke detector. A short time later the occupant again called the front desk to report that she and her bedridden mother were trapped in their room by the fire. Both room occupants perished in the fire with the official cause of death being listed as carbon monoxide intoxication for each victim. It was less than 20 feet from their room door to the outside of the building.

After notifying the fire department via the 9-1-1 emergency telephone number, the desk clerk called a hotel maintenance employee who lived in Room 224, just a short distance away. Although the employee responded promptly, the fire conditions in the breezeway outside Room 222 were impassable due to the flames.

The Adams County Communication Center (ADCOM) records indicate receiving the 9-1-1 telephone call at 2:33 a.m. and a second telephone call at 2:39 a.m. North Metro was dispatched at 2:35 a.m., sending four engines, an ambulance and a battalion chief; units were reported enroute within two minutes. Thornton police units were enroute to the fire even before the fire department was dispatched; their first cars were reporting flames visible from the outside on arrival at 2:39 a.m.

Engine 67 located approximately 1.8 miles away was the first fire unit on the scene several seconds before 2:40 a.m. The four person crew positioned the apparatus near the fire hydrant at the southwest corner of the front wing and initially assisted with evacuation of Room 211. The crew then stretched a 1-3/4-inch line to the base of the stairway in the southwest corner.

Battalion Chief 61 requested a working fire tone at 2:40:30 a.m. and a second alarm at 2:40:44 a.m. Unfortunately, due to the extensive radio traffic and activity at ADCOM, the chief's second alarm request was not recognized. The command post was established in the parking lot in front of the building with the battalion chief assuming the incident command. When the Incident Commander requested the response status on the second alarm units at 2:54 a.m., ADCOM advised that none had been dispatched. About 2:57 a.m. the second alarm was recognized and dispatched. This alarm brought four more engines and a truck to the scene.

Engine 62 was the second engine on the scene just before 2:47 a.m. accompanied by Ambulance 62. The three member engine company assisted Engine 67 with rescue and advance of the hoseline. The ambulance crew established a triage area in the front parking lot and requested additional resources from its headquarters. Eventually five AMR ambulances plus supervisory personnel responded to the fire scene. The search of Rooms 201 to 210 was completed by 2:52 a.m. and this area declared "all clear" by Engine 62. Attention now focused on extinguishing the fire at the base of the stairs at the west end of the front wing. (See Figure 3 for diagram)

Engine 66 arrived at the scene at 2:44:36 a.m. and the Incident Commander directed them to the west side of the property. The three member crew's initial operation was the evacuation of the west wing at the east end. A ground ladder was used to assist the occupants of Room 301. At arrival, fire was exiting at the second and third floor levels through the breezeway glass and aluminum enclosures. These had failed completely and the second floor breezeway area was heavily involved and the third floor level was nearly the same.

Engine 63 arrived about four minutes after Engine 66 and its three member crew was assigned to assist Engine 66. It reverse laid a 5-inch supply line from 66 to the hydrant at the north end of the property. A 1-3/4-inch line was advanced by the combined crews as the hydrant water supply was established. About 3:08 a.m. these crews found the first victim and removed him from Room 220. The victim was unconscious, not breathing, and did not have a pulse. The fire in the second floor breezeway and passageway was still burning in several locations.

Most of the second alarm units were enroute by 3:04 a.m. or before. Three of the engines and the truck were on scene by 3:10 a.m. and the fourth engine by 3:14 a.m. Engine 68's three member crew was assigned to the patio area at the inside corner of the junction between the front and west wings. They searched Rooms 101 to 106, 212 to 218, and the second and third floor west wings closest to the fire. Afterwards, they advanced a 2-1/2-inch line wyed into a 1-3/4-inch line to the patio area from Engine 66.

Truck 51 was positioned at the west end of the front wing and raised their ladder to the roof. The six member crew ventilated the roof over the stairway at the west end of the front wing and the stairway at the south end of the west wing. Assistance with this assignment was provided by the three member crew from Engine 33. The two stairways were ventilated about 3:30 a.m. and Command assigned the truck company to ventilate over the utility chase in the front wing. Engine 33 was given another assignment.

While the roof venting was underway, search of the west wing third floor rooms away from the fire continued with the assistance of the Thornton Police. Engine 63 and 66 crews were also searching Rooms 220, 221, and 222 off the still partially burning breezeway. A second fire victim was located in Room 220 and removed to Engine 66 at 3:39 a.m. Two more victims were located at 3:53 a.m. in Room 222. All victims were unconscious, not breathing, and did not have a pulse. The fifth victim, who was badly burned, was not located until later. His remains were found near the breezeway's second floor west enclosure wall in the debris. The coroner was requested to the scene at 3:51 a.m.

Engine 65 with a four member crew arrived at the scene at 3:14 a.m. and, along with Engine 33, were assigned to hand stretch a 2-1/2-inch line from Engine 67 to the east end of the front wing.

This operation started about 3:30 a.m. Reports from Engine 33 at about 4:00 a.m. indicate an immediate need for ventilation in this area and at 4:10 a.m. they reported that the fire in the utility chase had extended to the east end of the front wing. They requested additional assistance at this time.

Incident Command had also been requesting additional resources. The Red Cross was requested to help with the displaced occupants around 3:00 a.m. Buses were requested from the Regional transportation authority for both occupant protection and later for firefighter rehabilitation. No official third alarm was ordered; rather, requests for specific equipment were made. At 3:35 a.m., two engines were requested from Westminster Fire Department. Shortly before 4:00 a.m., breathing air supply and spare bottles were requested from North Washington Fire Protection District.

Chief 51 was supervising the roof sector and the ventilation being done over the utility chase and reported that progress was being made at 3:46 a.m. Unfortunately, about 3:52 a.m. two firefighters broke part way through the roof and one injured a knee sufficiently to be transported to the hospital for treatment. At this time all firefighters were ordered off of the lightweight roof. Fire through the roof near the southwest corner of the front wing was reported at 3:58 a.m.

Change to Defense Mode

Tactical operations then changed from an offensive mode to a defensive one. About the same time the fire conditions were deteriorating, Incident Command was transferred to North Metro's Deputy Chief of operations. Command instructed Truck 51 to establish a ladder pipe operation at its position and two 2-1/2-inch lines were stretched from Engine 67. The "B" shift was recalled to their stations at 4:13 a.m. and two engines were requested from Southwest Adams County shortly after. (See Figure 4 for diagram)

Westminster Engine 2 did a forward 5 inch hose lay from the hydrant at Acoma Street and 84th Avenue to near the southeast corner of the front wing. Westminster Engine 6 set up their TeleSquirt on the south side of the front wing using a supply from Engine 2. Both operations were completed and the elevated master stream operating before 4:30 a.m.

As crews became available, lines were advanced from Engine 2 to the front wing's second floor east end. Two 1-3/4-inch lines were placed near Rooms 201 and 212 respectively. Another 2-1/2-inch line was brought to the east end of the utility chase to back up the one placed into operation by crews from Engines 33 and 65.

Between 4:30 and 5:00 a.m., the main body of active fire was being contained in the front wing by master streams and handlines. Sufficient fireground strength was available that just before 5:00 a.m. Command advised sectors that crew rotation and rehabilitation was possible. Although not officially declared, the fire was effectively under control at this time.

DAMAGE AND CAUSE ASSESSMENT

While the fire was essentially contained to the front wing, it was burning within the entire length of the utility chase. The fire broke out of the chase at several locations on the first and second floors. Inside the utility chase, the exposed 2 x 4-inch wood studs were heavily damaged the entire length and the studs near the west end were totally consumed by the fire. This resulted in the roof collapsing over almost the entire length of the chase. Partial roof collapse also occurred over Rooms 219 to 222 and at the rear (by the utility chase) of Rooms 208 to 211. The fire entered several rooms on the second floor, consuming most of the rooms' contents. Yet adjacent rooms were not heavily damaged.

On the first floor north side, the fire followed along the wood construction in the separator walls between rooms, consuming the 2 x 4-inch wood studs from the inside out.

The point of origin is believed to be a first floor storage closet at the stairway located at the west end of the front wing. The closet was formed using the space under and adjacent to the stairway. The metal stairway with concrete filled steps and landing was enclosed on the underside with 5/8-inch drywall. The hollow concrete block stairway walls were also covered with furred out drywall. Where walls were needed to complete the closet enclosure, 2 x 4-inch wood studs were erected and covered with 5/8 inch drywall on each side.

The closet was nearly filled with rolls of carpet and foam padding at the time of the fire. The wood door of the closet at the beginning of the fire is believed to have been partially open. This allowed combustion air into the closet and the heat, smoke, and flames to quickly rise up the adjacent stair opening to the second floor corridor and breezeway. The storage closet was also adjacent to the utility chase which allowed smoke, and eventually fire, to enter the chase and then extend to the guest rooms.

BUILDING AND FIRE PREVENTION CODES

The hotel was constructed in 1963 and would have had to comply with Adams County building codes and ordinances. The applicable building and fire prevention codes in effect at the time of construction and could not be identified for the report. The hotel property was later annexed in to the City of Thornton which provided for building code enforcement and issued permits for the hotel's operation at the time of the fire. Thornton adopted and was enforcing the 1991 Uniform Building Code and Fire Code when the fire occurred.

The North Metro Fire Rescue Authority was enforcing the 1991 Uniform Fire Code during its inspections at the hotel. The hotel had received regular fire inspections since 1980, the longest period for which these records were available. The property was considered a target hazard under both the pre-incident plan and the fire prevention program. The hotel was often visited twice a year and more frequently if compliance with serious deficiencies was required. These inspection records indicate that at the time of the fire, the hotel did not have outstanding any significant issues which would have influenced the fire's outcome.

A major fire protection achievement was the installation of automatic sprinklers in the basement banquet room during 1993 and 1994. One measure assisting the Authority's effort to get automatic sprinklers installed was the hotel's need to have a permit to operate a place of assembly such as the banquet room, restaurant, and lounge. The permit had specific expiration dates and needed to be renewed to continue operations legally. The permit process was controlled by North Metro Fire Rescue Authority.

The following analysis uses the existing hotel provisions of the 1994 edition of NFPA Standards No. 101, *The Life Safety Code®*. The City of Thornton and the North Metro Fire Rescue Authority did not adopt or enforce NFPA 101 and there was no legal requirement for the hotel to comply with this code. NFPA 101 is being used in the analysis because of its wider availability as a reference tool. Because of the multiple street levels for exit discharge and the need for compliance with special conditions, the analysis results are subject to local authority judgment and acceptance.

A fire alarm system would be required for hotels except for buildings not more than three stories high where each guest room has exterior exit access. The East Wing had such a system and except for the few rooms at the 200 level, the balance of the hotel met the exception. Four of the fatalities occurred in rooms which did not meet the exception and the fifth fatality was an occupant of these rooms.

When guest rooms are arranged along an interior corridor, the walls separating the room from the corridor are required by code to have at least a 30-minute fire resistance rating. The guest room doors are to be at least 1-3/4 inch thick solid bonded wood core and self-closing operation. In addition, no louvers or transfer grills are permitted in the separation walls or the doors as these would provide a means for smoke spread into the guest rooms. While the guest room doors complied with the code, the separation wall had a large glass window which would not comply. Each room had a self contained combination heating and air-conditioning unit under the window and these units could use air from the corridor which is similar to a transfer grill. The window and the heating and air conditioning unit raise a question about Rooms 219 and 222 complying with the code's corridor separation provisions.

The original hotel construction had the stairways and breezeways covered but outside the building. The original arrangement would have allowed smoke and heat to vent and not be contained where it could rapidly block guest room exiting. As constructed, Rooms 219 and 222 would have complied with Life Safety Code provisions for fire alarm and guest room separation from corridors. The glass and metal walls and glass doors constructed at the ends changed open exterior stairs into interior stairs and created an interior corridor. An analysis of the proposed enclosure construction should have identified the need for additional fire protection features for the stairs and guest rooms adjourning the breezeway.

Unprotected vertical openings, such as stairways and utility chase, connecting up to three stories are permitted when certain conditions are met. In the author's opinion not all of these conditions were complied with to permit the open stairways and utility chase. Specifically not met was the occupants' ability to recognize a fire in any part of the connected space prior to the time the fire becomes a hazard to them. Also, the requirement that the contents in the space be low hazard or ordinary hazard protected by automatic sprinklers was not met. The carpet and other materials in storage would probably not be considered low hazard and the print shop contents would not be considered low hazard either. These spaces communicated with the stairway and the utility chase and were not protected by automatic sprinklers.

PAST FIRE EXPERIENCE

During 1995 and 1996, records identified four fire incidents associated with the property. One incident was in a car that, according to Thornton Police, had been parked at the property for a long time with expired license plates. This fire was confined to the car's interior and principally involved the seating. Another fire involved hotel maintenance workers soldering copper pipes which ignited a smoldering fire in the utility chase. Once the source of the smoking fire was located, it was promptly extinguished with minimal damage.

Two other fires involved the guest room wings. In 1996, a fire involved some bedding which had been placed outside one of the third floor guest rooms. This fire was found in the early stage and was extinguished with a portable fire extinguisher before the first engine arrived. In 1995, a fire, described in the report as of suspicious origin, occurred in one of the second floor guest rooms in the same wing but on the south side of the building. The involved guest room exited directly outside so the heat and smoke were not trapped. The fire was principally confined to the room or origin with smoke and water damage to the exterior walkway and two adjacent rooms. There were no injuries reported in either fire.

FIREGROUND COMMUNICATIONS

The ADCOM operator who was handling this incident was new and not experienced with all the procedures or the amount of activity a major incident can generate. Battalion Chief 61 requested a second alarm while enroute to the call and before fire department units were on the scene. The ADCOM operator did not fully understand the message. Rather than a second alarm, the dispatch tones for a working fire were activated again which notified all of the North Metro Fire Stations. A delay of about 14 minutes transpired before the resources for a second alarm were dispatched. The delay could have been longer as the recognition did not occur until the Incident Commander requested a status report on the arrival of the second alarm resources. Monitoring for acknowledgment of requests or listening for the actual dispatching being done is an important function of the communication process, which would include both voice and data modes of transmission.

Firefighters, company officers, and sector commanders encountered the characteristic communication difficulties when self-contained breathing apparatus (SCBA) is being used. The problems involved both verbal communication within company operations and the transmission of radio messages.

ACCOUNTABILITY

North Metro Fire Rescue Authority uses a dual passport system where company members are listed on two separate places. One passport stays in the apparatus and the other travels with the company officer. Passports can then be handed off to group or sector control/accountability positions. The passport system and usual fireground operations anticipate that companies work as a unit on the fireground. In this fire, some companies were split because of the conditions. Part of the company was assigned to search rooms and the remaining members strengthened another company involved in the fire attack. Accounting for all firefighters under these unusual conditions is a challenge to the company officer and the system. One means could be to handoff the detached members to the other company officer for accountability purposes.

Firefighter accountability should not be overlooked during small fire incidents or in the beginning of a potentially large incident. The need to establish an accountability system grows in importance with multiple sector operations. In this fire, early fireground demands for simultaneous rescue and suppression operations raised the possibility for splitting companies. Accountability can be difficult at this stage because sector officers may not be specifically identified and attention is often focused on other pressing matters. The fireground may also have police officers assisting with rescue and evacuation complicating the complete knowledge of all rescuers operating in the vicinity of the fire.

During this fire, first arriving companies were involved with multiple fireground tasks and were sometimes grouped into operating units assembled from more than one company. No problems were reported either with the operation of the accountability system or with firefighters becoming missing during this incident.

FIRE INVESTIGATION

The Thornton Police began to interview hotel occupants regarding their observations even while the fire suppression was still underway. The North Metro Division Chief/Fire Marshal was requested to the scene early in the fire to organize the fire investigation. The fire marshal also served as the public information officer during both the fire suppression phase and during the subsequent investigation phase.

The fire department received assistance first from other mutual aid department investigators and the Thornton Police Department. The Bureau of Alcohol, Tobacco & Firearms (BATF) Denver Office provided several agents initially. The BATF National Response Team was activated on Monday and began arriving around noon on Tuesday. In this incident, the BATF Team brought 14 members and a dog and the Denver BATF office supplied five agents including the case agent.

For the scene investigation, the BATF Team consisted of a supervisory staff, evidence technician, laboratory specialist, K-9 specialist and K-9, photograph specialist, schematic specialist, and a certified fire investigator. The scene investigators were assisted by an organizational staff, interview specialist, and technical support personnel.

The Team integrated with the local fire and police members into a common working group. The overall team coordinator was North Metro Division Chief/Fire Marshal with assistance from the BATF Case Agent and the Lead Detective from the Thornton Police Department. Because of the number of occupants, other witnesses and investigation leads, an important element BATF assisted with was a computer-based cross-reference system to organize the data being collected by the team. The system also allowed the preparation of a fire time line for data correlation as it was assembled from witness statements and fire scene evidence.

The fire investigation continued on a steady but perhaps not a continuous basis for over three months. This included members from North Metro, Thornton Police, and the Denver office of BATF. The cause of the fire was determined to be incendiary and the origin being the 100-level storage room at the west end of the front wing. A monetary reward has been offered and advertised in the print, radio, and television media. Funds were provided by the hotel owner and the State Arson Hotline Association.

LESSONS LEARNED

1. **Fires in residential occupancies with center loaded corridors continue to block exit and escape paths resulting in injuries and fatalities.**

   In this case, the fire trapped occupants in their rooms and there was no alternate escape path available. Nor was there an alternate source of smoke-free air to permit the occupants to use their rooms as places of refuge. The fire and smoke was infiltrating the room from both the front and from the utility chase at the rear. All of the fatalities occurred in or came from rooms which were in the breezeway. All of the occupants of rooms that exited to the exterior open walkways and balconies were able to escape without significant injury, even after delayed fire awareness due to the lack of a building-wide fire alarm.

2. **The hotel did not have a building-wide occupant notification fire alarm system that included the front and west wings.**

   The guest rooms contained battery operated single station smoke detectors which activate individually from the conditions in each room. These units warned room occupants of a fire that originated in their room or smoke entering their room from an adjacent space. As long as their means of escape was not blocked by the same fire, the occupants could be expected to react to the alarm and leave before life-threatening conditions develop. Occupants of rooms with direct outside exits escaped the fire whether notified by their single station smoke detectors or by people knocking on their door. Life Safety Code provisions for existing hotels with rooms exiting directly outside do not require a fire evacuation alarm for such areas. Battery operated smoke detectors, and even most hardwired single station smoke detectors, are not intended for general evacuation alarms or notification of guest room occupants of a fire in another part of the occupancy. Manual door-to-door alarm notification is often needed in these conditions and the resources to accomplish such tasks over and above the suppression force should be included in the pre-incident plan. Some success has been reported in other incidents of using apparatus sirens and horns to notify occupants of a fire. Even in structures with code compliant fire alarm systems, physical searches of the threatened units must be undertaken quickly.

3. **The fire was not detected and fire department not notified until it was of substantial size and was rapidly spreading.**

The lack of automatic fire detection in the area of origin combined with the time of night and the storage room's secluded location made early occupant fire detection unlikely. By the time of fire detection and notification, occupants of Rooms 220 and 222 were unable to evacuate before their means of escape became blocked. Where interior corridors are part of the exit path, prompt fire detection and immediate occupant notification is essential to effect their safe evacuation. Occupants' safe evacuation depends upon their quick response to the alarm signal and that they are capable of self preservation. Alternate occupant safety can often be provided if fire growth is stopped and the generation of combustion products controlled. Quickly attacking the fire manually or automatically are means to accomplish these tasks.

4. **The utility chase between rooms and floors provided a rapid and easy means for the spread of combustion products initially and the fire later.**

The utility chase was not smoke or fire stopped either vertically or horizontally within the chase. In addition, the chase was not separated by fire resistance rated construction from the guest rooms. Even the first floor hollow concrete block walls were not completely separated from combustible concealed spaces in the guest rooms. On the second floor, the wood studs were exposed in the utility chase providing a fuel source for the fire to move horizontally inside. As the wood studs failed, or the fire found openings in the gypsum board wall, the fire was able to enter the guest rooms, especially where fuel was available in the room near the utility chase wall. Several guest rooms were gutted by fire in this manner. The recognition of combustible concealed spaces is an important part of the pre-incident plant process. Fire companies or fire protection staff need to carefully investigate the building's construction and physical arrangement to identify potential concealed spaces. The more difficult task is developing and implementing actions to mitigate construction deficiencies such as lack of smoke and fire stopping, exposed combustible materials or lack of fire resistance rated separations.

5. **Requesting additional alarms is usually a better means of obtaining and staging fireground resources during fires in large residential occupancies such as this.**

Although there was confusion on the part of the dispatcher resulting in delay for the second alarm, the Incident Commander's requests were appropriate for the occupancy and conditions being reported. Officially a third alarm was never requested by the Incident Commander yet by the end of the fire, the number of companies on the fireground was reportedly equal to such a response. By requesting equipment through additional alarms rather than by individual companies, the Incident Commander provides for less radio traffic and an orderly progression of resources to the fire scene. Another benefit may involved companies making predetermined moves to fill vacated quarters, restoring a level of emergency service in these areas.

Another consideration in developing tactical plans for fires in large residential occupancies is the need to balance fireground resources between search and rescue and fire suppression. The limited number of fighters initially on the fireground makes it a challenge but both must be done simultaneously because concentrating on one can result in neither being accomplished successfully. The fire spread needs to be at least slowed to allow companies the time to complete search and rescue operations. Yet those residents immediately endangered by the fire must be rapidly evacuated and some may require assistance in doing so. In some structures, even residents remote from the actual fire can be overcome from the smoke and the large quantity of carbon monoxide generated during fire suppression. In the competition between search and rescue and fire suppression, extinguishing the fire usually helps reduce the life loss hazard substantially.

6. **Fire resistance rated storage rooms do not always provide a reliable method of protecting the means of egress from a fire originating in the room.**

The rolled carpet was stored in a room constructed under and adjacent to the noncombustible stairs connecting the 100 level to the 200 level. The storage room was constructed with materials that provided roughly a one-hour fire rated enclosure except that the door may have been a solid core bonded wood core unit rather than a rated fire door. The evidence is not clear regarding the presence of a self-closing device on this door. It is believed that the door remained at least partially open during the fire and was mostly consumed during the fire. Even though this was an incendiary fire and the perpetrator could have blocked the door open, accidental fires have occurred with similar circumstances. Storage rooms under and adjacent to noncombustible stairways are common and an uncontrolled fire represents a potentially exit blocking exposure.

7. **Unenclosed stairway openings provided an easy avenue for smoke, heat, and fire to spread through three levels of the building and trap the occupants whose guest room doors opened into this path.**

The original hotel construction likely did not have walls and doors that enclosed the second and third floor breezeways of the stairway/walkway area at the west end of the front wing. This means that all guest rooms opened directly to the exterior including those whose occupants died in the fire. The stairways, while covered, were also open to the exterior of the building. Sometime later enclosures were erected at the first, second, and third floors to protect these spaces from the weather. The enclosures were principally constructed of glass doors and glass walls in metal (believed to be aluminum) frames. Additional construction included storage rooms under and adjacent to the stairways at all three levels. The two stairways were not isolated from the walkways or the breezeways including the one that the three second-floor guest rooms opened into. The open stairways allowed the now-enclosed walkways and breezeways to rapidly fill with smoke and heat. Eventually these areas flashed over from the fire products.

8. **Quickly identifying and locating occupied guest rooms is important during the early evacuation and the initial search and rescue phase of the fire.**

Guest room occupancy records are essential and room number diagrams are important to direct companies to those occupied rooms immediately in danger. North Metro Fire Rescue Authority's pre-incident plan included a room number diagram and the desk clerk brought the hotel's records. Unfortunately, it was not immediately discovered that shortly before the fire the hotel had revised parts of the room numbering system. The pre-incident plan diagram was not accurate and some searches of unoccupied rooms were made during the initial phase before the change was discovered. Fortunately, the hotel room occupancy records were available to the Incident Commander. As part of a hotel pre-incident plan, the hotel management should be educated about the importance of removing occupant records and providing them to the fire department in the event of an emergency. In this incident the hotel lobby was not seriously threatened by the fire and accessing the records was simple.

APPENDIX A

Incident Diagrams

Appendix A (continued)

Appendix A (continued)

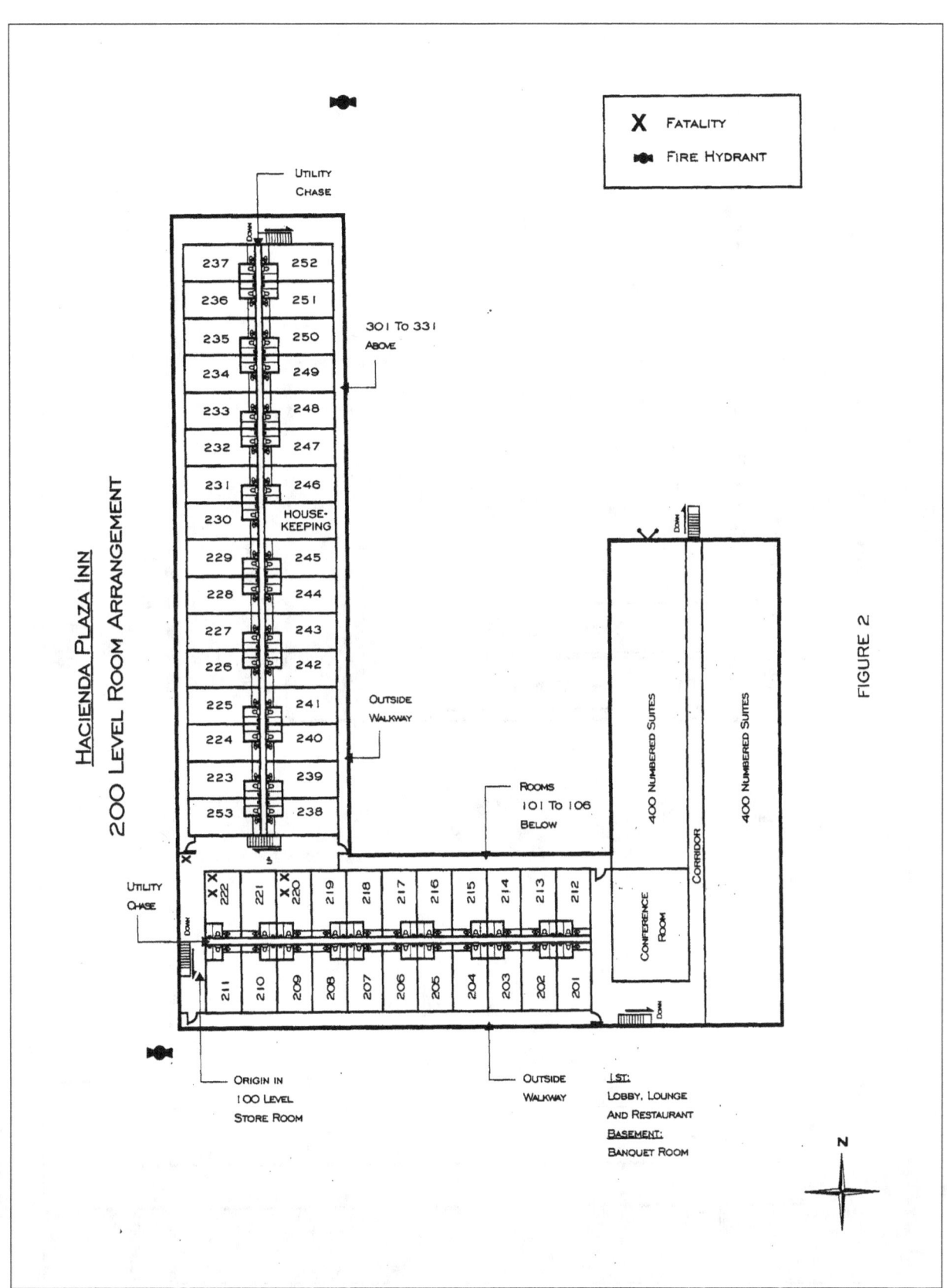

Appendix A (continued)

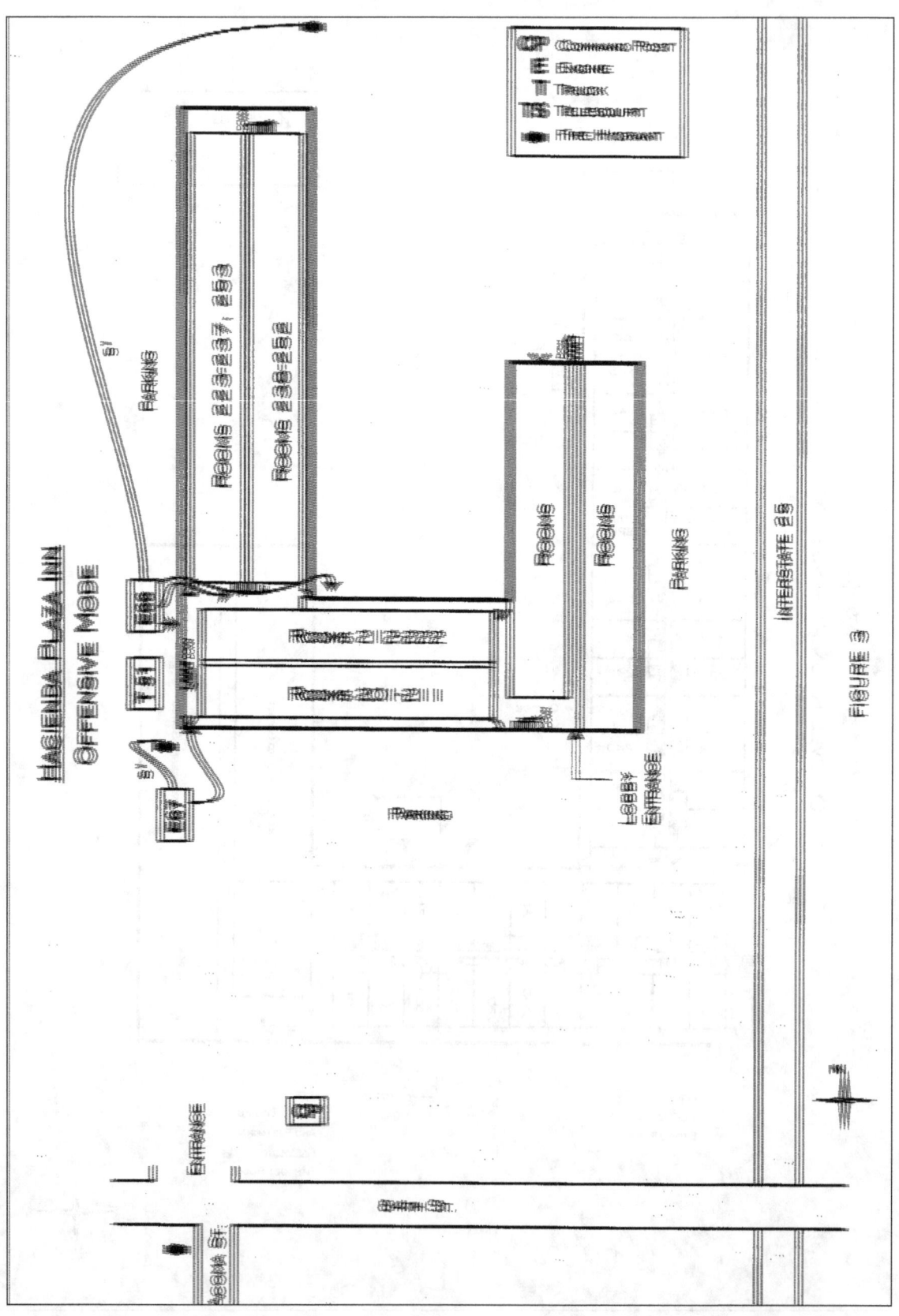

Appendix A (continued)

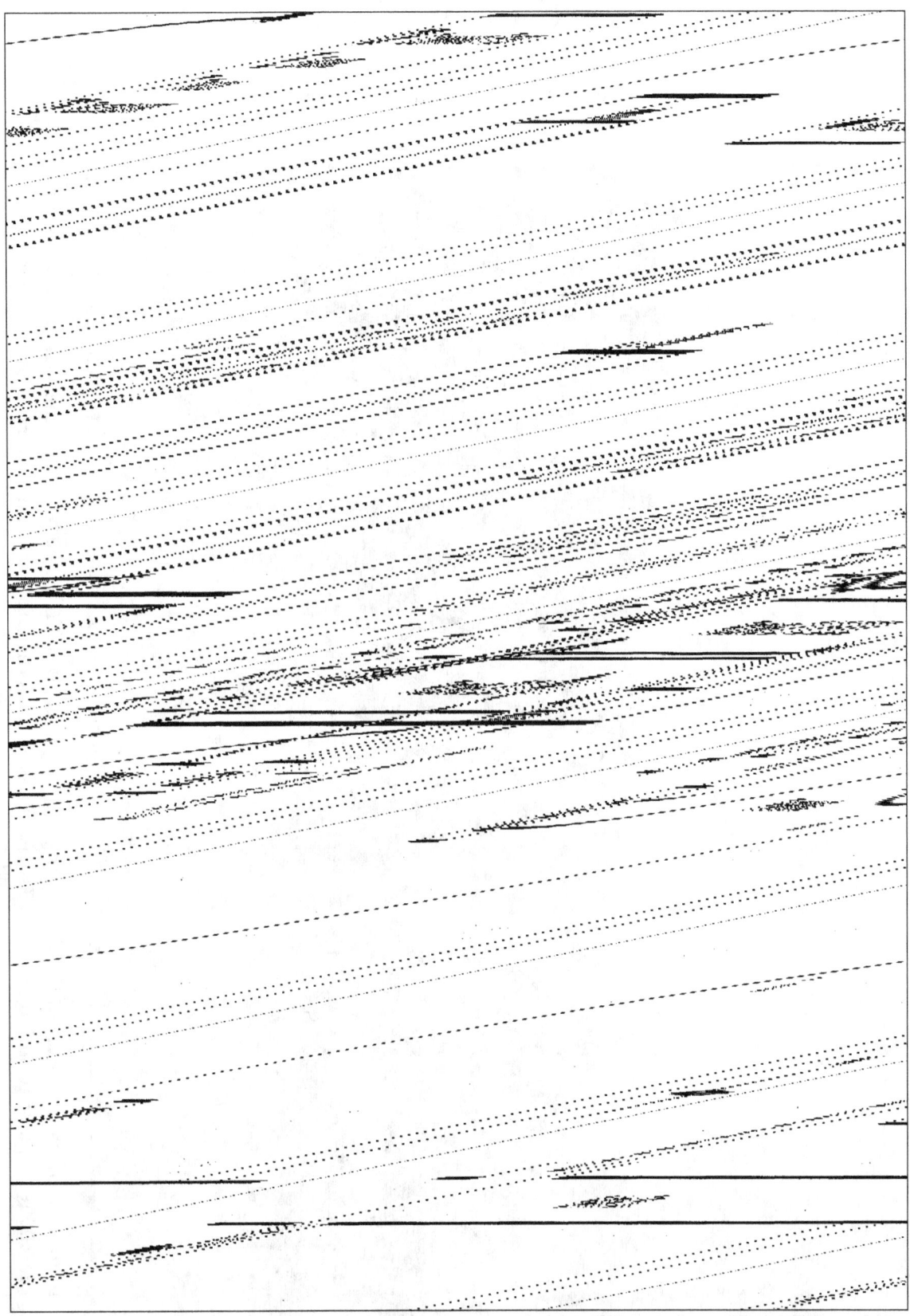

APPENDIX B

Photographs

All photographs were taken by Thomas Miller.

Photo 1. Southwest corner of the front wing looking northwest. The 100 level, commercial and offices, and the 200 level guest rooms are shown. The pine tree shields the entrance to the stairway and storage closet which was the point of origin.

Appendix B (continued)

Photo 2. The west side of the property looking north. This is the 200 and 300 level guest rooms of the west wing and the end of the front wing. Engine 67 used the fire hydrant to the right of the light pole.

Appendix B (continued)

Photo 3. The inside corner of the junction between the front and west wings; part of the U shape. This illustrates all three levels of the property and the grade change.

Appendix B (continued)

Photo 4. Typical guest room ceiling, bathroom, and sink of the utility chase wall.

Appendix B (continued)

Photo 5. The interior of the utility chase between rooms showing the second floor wall construction and the wood beams supporting the roof.

Appendix B (continued)

Photo 6. Roof over the front wing. The collapsed area is above the utility chase. The roof construction of Tectum board, built-up roofing, and gravel topping is illustrated.

Appendix B (continued)

Photo 7. The west end of the utility chase in the front wing. Room 222 is to the left and Room 211 to the right. The second floor utility chase walls were burned away.

Appendix B (continued)

Photo 8. The 200 level breezeway area outside of Rooms 220 to 222 looking west. The open stairway to the 300 level is on the right. Under the stairs is another storage room whose wood stud and gypsum board walls were destroyed by the fire.

Appendix B (continued)

Photo 9. A separation wall between 100 level guest rooms that was burned from the inside out. The parallel wood beams formed the top of the wall and the communication to the utility chase is at the back.

Appendix B (continued)

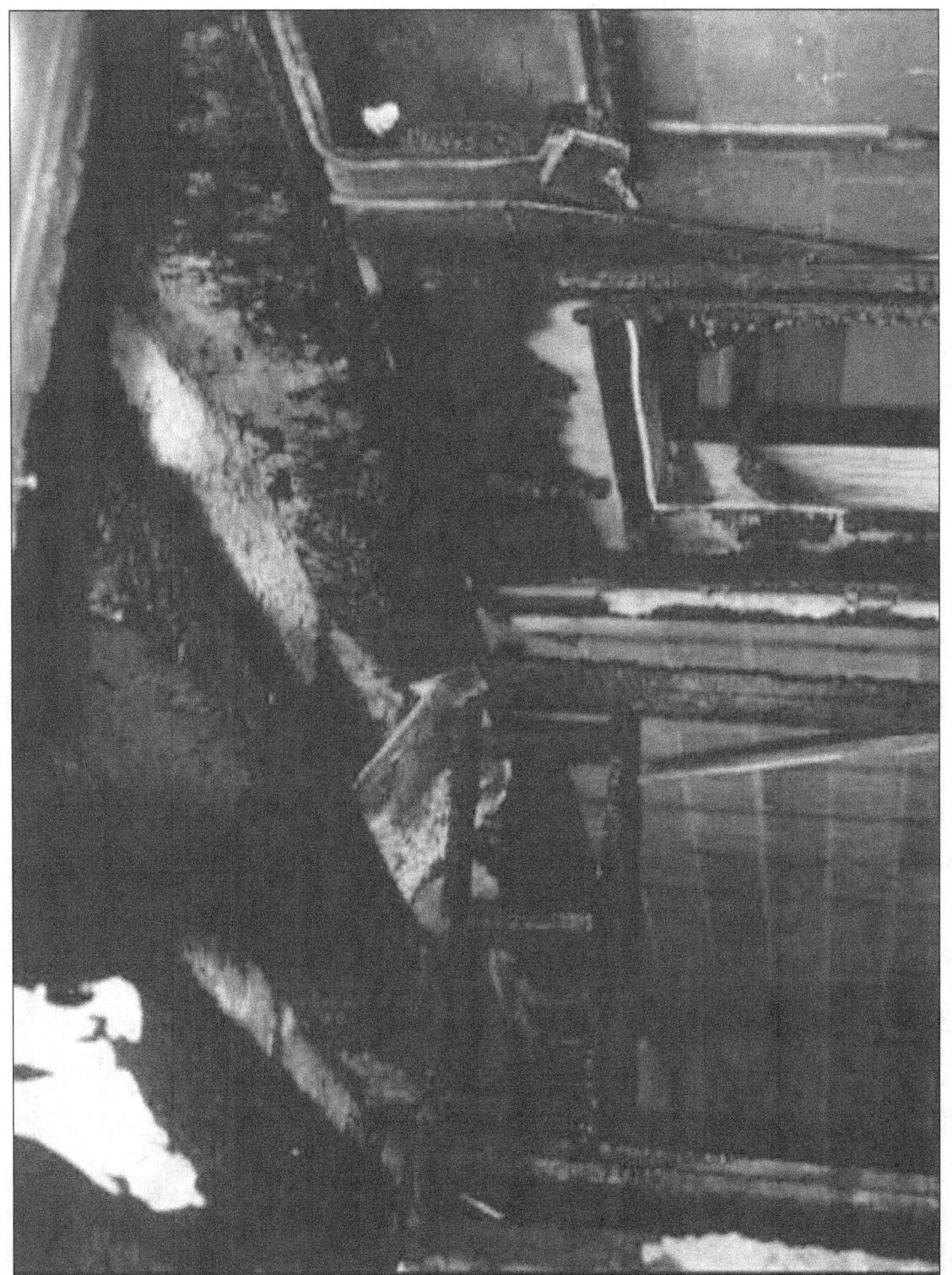

Photo 10. The top of the door frame for the 100 level closet determined to be the point of origin. The open stairway to the 200 level is to the top left off the photograph.

Appendix B (continued)

Photo 11. The top of the open stairway from the 100 level looking south. A glass in aluminum frame wall enclosed this area just before the partial height brick wall in the background. Room 222 is to the left.